# The Navigators' Guide
# to Hyperbolic Navigation

## H. NEVILLE DAVIES, A.M.Brit.I.R.E., M.I.N.

(Author of "The Navigator's Guide to Radar")

# FOREWORD

MANY means of obtaining navigational information by radio exist; some old, but the majority of them quite new in development, and all of them producing the information in the form of a line of position. Most of these systems involve three signal points. Those with two fixed receiving points and a mobile transmitting point are classed as direction-finding systems. Those with two fixed transmission points and one mobile reception point are navigational systems. The former has long been familiar to the modern navigator, but advances in knowledge of wave propagation together with improvements in design of equipment have increased its accuracy, so much so that the appearance of many new aids has not lessened the value of the familiar D.F. apparatus as a navigational aid.

However it is not proposed to deal with the Direction-Finding System, which nevertheless remains the only system affording facilities for determining the position of a ship in distress, but to deal with the newer systems now being introduced into the Merchant Navy using the new art of Hyperbolic Navigation.

In dealing with these modern devices two important points must be borne in mind. Firstly, all navigators who use these devices as a sole reliance run the risk of forgetting the ordinary technique of their trade. And then, when a valve conks out, or a resistance burns out, or power sources fail—or they get into areas outside the range—they're out of luck. Like the rest of the navigational aids they're intended to be just that—an aid. As such they've proved to be good ones.

Secondly, it must be remembered that it would be inappropriate to express a definite opinion of all the radio aids to marine navigation which have been demonstrated and discussed, until they have been operated and tested under all conditions by ships' navigating officers, who must be the ultimate judges of the fruits of scientific research.

# CONTENTS

# CHAPTER I.

## FUNDAMENTALS OF HYPERBOLIC NAVIGATION.

THE development of the art of Hyperbolic Navigation progressed from the basic principles being used in the technique of radar. With the development of the Cathode Ray Tube as a timing device of great accuracy coupled with the already accurately measured speed of travel of wireless waves through the air, it was apparent that an instrument could be constructed for carriage on board ship which could measure the time it would take a signal to arrive at the vessel from a known transmitting station ashore. Knowing the speed of travel of the signal this time would give an indication of the distance from the transmitter.

The problem, of course, was the high speed of wireless waves, which travel at the same rate as light, i.e. 186,000 miles per second or, in other words, a sea mile every 6 millionths of a second. However, a Cathode Ray Tube enables us to measure time to millionths of a second so a stopwatch was at hand.

This system had an immediate application to marine navigation when used in the following way. Two shore stations are used and they each transmit a signal, in the form of a short 'pulse, at an identical instant of time. In the receiver on board ship the pulse signals are received and displayed in some way, which for the purpose of explanation we will presume is by means of visual presentation on a Cathode Ray Tube. Obviously, if these two signals arrive on board at exactly the same time the signals have travelled the same distance, and the vessel must be equidistant from each of the two stations, which position could be exactly in the centre of a line joining the two transmitting stations. (See Fig. 1.) But there would be an infinite number of these positions all equidistant from the two transmitters, and all of them lying on a line exactly at right angles from the centre of the line joining the two stations and on both sides of this line.

In other words one could simply take a chart, plot the exact positions of the two shore stations,

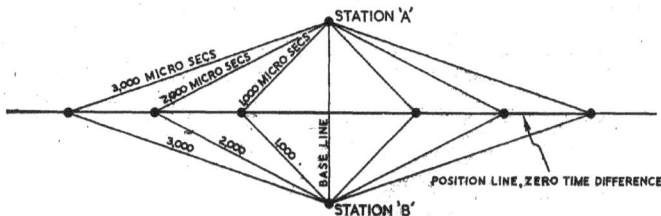

FIG. 1.—Simple Pair of Transmitting Stations. Showing Equidistant Line with Zero Time Difference.

draw a line joining them and mark off a further line at right angles to the centre of the base line. Any point on this line is equidistant from the two stations and a radio signal from them would arrive at a ship anywhere on this line at exactly the same time. (Fig. 1).

A ship in some other position would receive the signals at different times, the nearer station signal would arrive first with the farther station signal

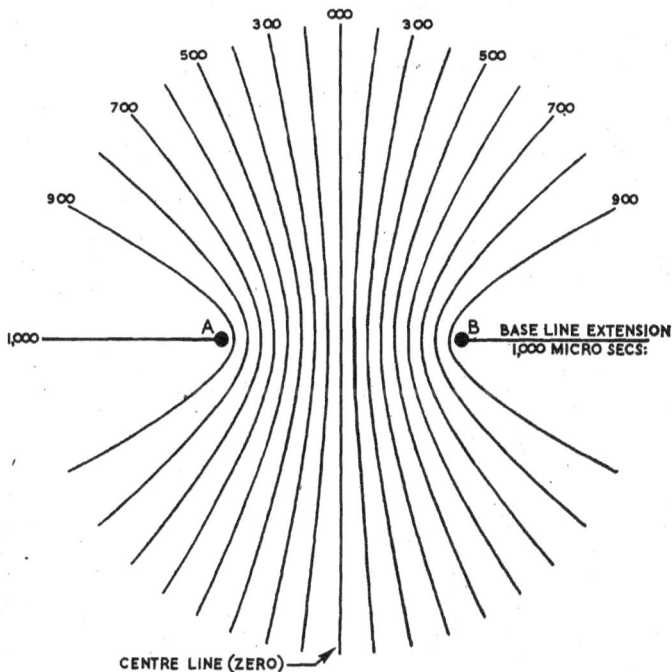

FIG. 2.—Lines of Constant Time Difference from a Simple Pair of Transmitting Stations. The Lines are Hyperbolae.

following. With the aid of our timing device the time difference of arrival can be measured. Again

11

there will be an infinite number of points in which a vessel is in a similar position relative to the two stations, although in an entirely different position geographically. A line can be drawn joining all these positions but it would no longer be a straight line but a hyperbola. (Fig. 2.)

These lines again can be drawn upon a chart mathematically with reference to the positions of the two transmitting stations.

With these two stations operating ashore fitted with a device afloat which will receive the signals and measure the time-difference of their arrivals on board, and armed with a chart on which the hyperbolic lines have been superimposed, we have a system which will give us speedy and accurate position lines at all times almost irrespective of weather conditions.

Of course, using this arrangement we only obtain a position line. To obtain a fix we require a further line. This can easily be done by providing another pair of stations ashore giving us an entirely different pattern of lines on the chart which will cut the first set of lines. All we require now is to get a position line from one pair and another from the second pair; where they cut is our fix. To obtain greater accuracy a third pair of stations can be operated giving still another line to aid in fixing.

12

The lines of constant time difference for each pair of stations are all pre-computed taking into account curvature and eccentricity of the earth, and other factors, and are made available to the navigator in the form of tables or special charts.

This is the idea behind all the hyperbolic systems, the arrangement of stations being called " Chains " and a standard chart with the hyperbolic position lines superimposed on it, is known as a " Lattice Chart ". The lines produced by the different pairs of stations are distinctively coloured on the chart and referred to as " Red Pair," " Green Pair," etc.

In actual operation the number of shore installations is cut down by using three stations to provide two lines. One station is known as the " Master " station, one of the others is a " Red Slave," and the third the " Green Slave ". The Master station signal and the Red Slave station signal being used for the Red Pair and the Master and Green Slave forming the Green Pair. (Fig. 3.)

Naturally the technical work in transmitters and receivers is considerable and the technique of measuring the time differences is not too simple.

However, the apparatus required on board is small, light, and easily fitted ; consisting simply of receiver, display and small aerial.

It will have been gathered from the above that Hyperbolic Navigation is entirely different from the

13

other means of radio navigation, i.e. Radio Direction-Finding and Radar. In this method we measure

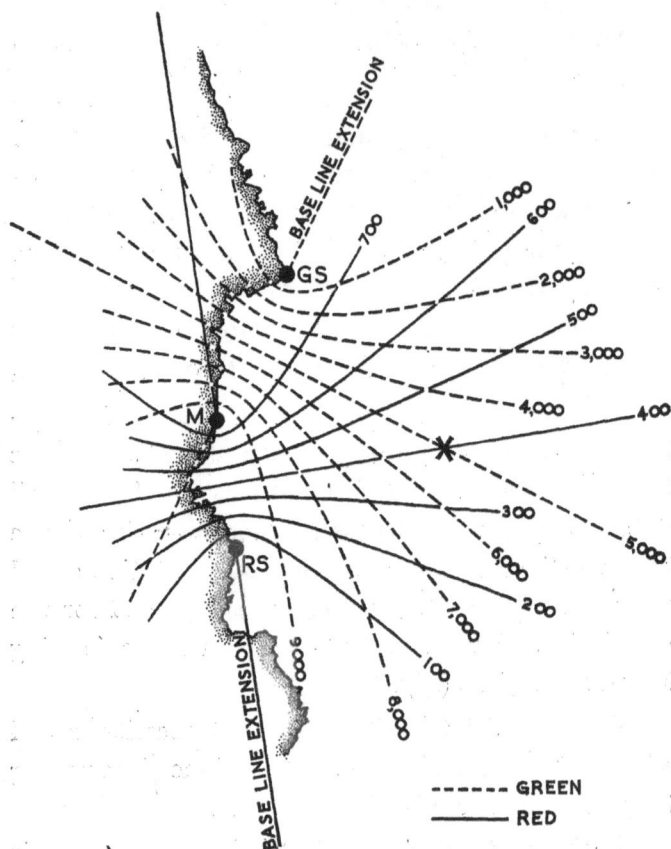

FIG. 3.—From 3 Stations Two Sets of Hyperbolic Lines can be Drawn on a Chart. M. is the Master Station, R.S. the Red Slave, G.S. the Green Slave. X. Indicates the Position of a Vessel whose Hyperbolic Navigator tells him he is on a Red 400 Line and also on a Green 5000. The Lines of course would be appropriately coloured on the Chart.

the time of arrival of radio waves whereas in Direction-Finding the *direction* of arrival of the waves is measured. The aerials used to receive the signals may thus be simple straight wires rather than loops or involved aerial systems. In Hyperbolic Navigation, unlike radar, no transmission takes place from the ship which simply receives radio signals from the shore transmitters.

However, as mentioned, some of the technique of time measurement has been borrowed from radar.

# CHAPTER II.

## LORAN.

THIS system as its name implies is a *Long Range Navigational* aid, employs the basic principles already outlined and is intended mainly for marine navigation and long flights over the sea.

It employs pairs of pulse-transmitting ground stations which are spaced up to 600 miles apart. The Master and Slave stations do not transmit at the identical moment, but the transmission is so arranged that the Master station radiates first and is followed by the Slave station after a fixed delay. This delay is allowed for in the receiver and the variation from this fixed period is of course, the time difference required.

The Navigator is, as always, primarily interested in the apparatus which he will be called upon to use and which supplies him with the information he requires for position-fixing. Loran signals are received aboard on a Loran receiver equipped with an ordinary aerial, and is basically similar to ordinary radio receivers but specially adapted to Loran. The output of the receiver does not go to a Loud-speaker but to a Loran Indicator which incorporates a Cathode Ray Tube and presents the signals visually.

The Indicator has two horizontal lines which appear on the screen when the set is switched on. In actual fact these are not lines at all, but simply a spot of light moving quickly across the face of the

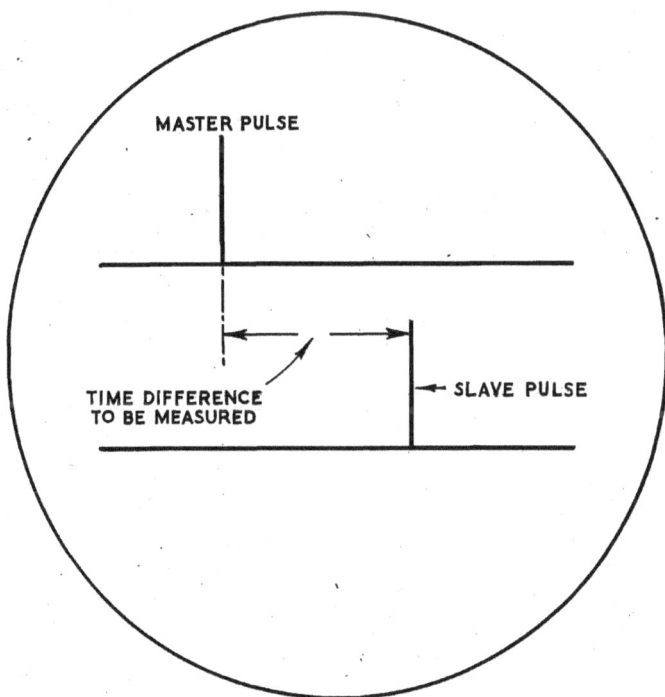

MASTER PULSE

TIME DIFFERENCE TO BE MEASURED

SLAVE PULSE

FIG. 4.—Simple Drawing of Cathode Ray Tube Showing How the Received Pulses are Indicated.

tube. Such is the speed of this spot, allied to the persistence of human vision that they appear as steady lines of light. The circuits are so arranged

17

that the Master pulse appears on the upper trace with the Slave signal on the lower. (Fig. 4.)

A control on the receiver allows the speed of the spot, across the tube face, to be varied, and its purpose is as follows. The Loran transmitters radiate on one fixed wavelength, to which the receivers are accurately tuned, but the different pairs radiate their pulses at different recurrence rates. Instead of having to tune the receiver to a particular wavelength the control on the receiver is simply put to the required position for the pair required, let us say the Red Pair, and the speed of the spot across the tube is automatically adjusted accordingly. If we now switch on the receiver the traces will appear on the tube and many pulses will be seen to be travelling along the traces from right to left. On closer observation it will be seen that one pulse on each trace will be stationary. These will be the required pulses from the Red Pair as the speed of the traces will be synchronised with the speed of transmission of the radiated pulses at the transmitter, so that the received pulses appear to be stationary whilst the pulses from other chains appear to move. This is similar to the ordinary film show where persistence of vision makes objects appear to stand still when they are in the same position in successive pictures, and appear to move when they are in slightly different positions in successive pictures.

When the receiver is switched on there may be

more than just two stationary pulses and by means of an other control the stationary pulses may be made to move along the trace. They are moved along until they take up a position similar to that shown in Fig. 4 i.e. the pulse on the top trace at the left-hand end of the top trace, with the other to its right on the lower trace.

These are the Master and Slave pulses and the distance between them is in effect the difference in time we require as shown on our " electronic watch."

The only part of the screen we now require is that portion containing the two pulses, so by moving another switch the picture on the tube is enlarged and only shows the required sections of the two traces containing the pulses.

We are now ready to measure the distance (i.e., time difference) between them.

This is effected by turning a knob on the face of the receiver which moves the Slave pulse (the one on the lower trace) along until it is exactly below the Master pulse. This knob is geared up to a counter which records the distance moved, not in terms of distance, but in time measured in microseconds. After this coarse adjustment a further control enlarges the picture still further so that the screen only contains the two pulses. A fine adjustment is then brought into use and finally a switch actually superimposes

19

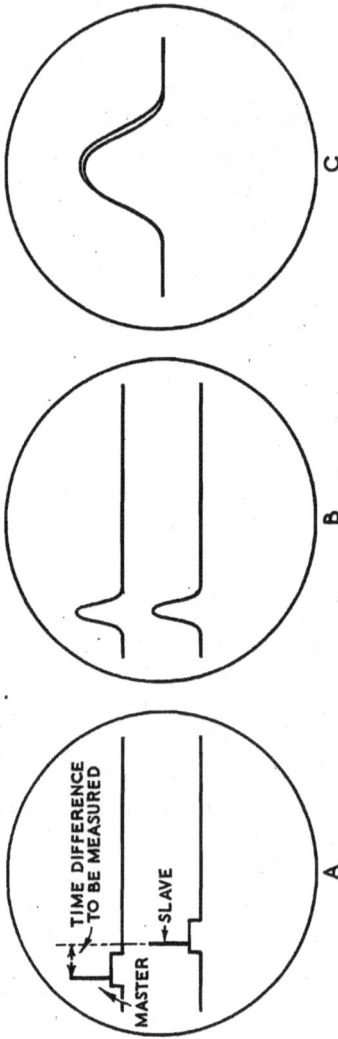

FIG. 5.—Showing How the Signals actually appear on the Loran Indicator. In A. the whole Traces are shown with the Two Pulses on Pedestals on the Traces. In B. the Pedestal containing the Slave Pulse has been moved along the Lower Trace until it is below the Master Pulse, and then the Picture enlarged until just the Pedestals are shown. In C. the Slave Pulse has been superimposed on top of the Master for accurate adjustments. All movements of the Slave Pulse being automatically recorded.

the Slave pulse on top of the Master, so that by using the fine control the pulses can be made to exactly coincide with one another. The counter should now indicate the exact time difference, which is the figure required. (Fig. 5.)

By reference to a Loran Lattice chart a position line bearing this time difference will be found. In order to get a fix the operation must be repeated using another pair of stations, i.e., the control altering the sweep time of the traces must be adjusted to the position in which a different pair of pulses appear stationary on them. It should be noticed that the two position lines are not taken simultaneously but one after another, with the time interval required to readjust the receiver. In practice it is usual to take, say, three readings of each pair and use the means of the results for obtaining the fix. With good reception conditions the observations can be made and the plotting completed in about 5 minutes, after some experience.

The wavelength used is such that the night-time range of this equipment is much greater than during the day, but much experience is required in matching the signals at night due to the unsteady behaviour of the pulses, due to what is known as "sky-wave" effect. The equipment required on board is light and easily fitted into any chart room and will provide service up to a range from the transmitters of 700 miles by day over sea, and 1,400 miles by night. The accuracy of a position line obtainable from a pair of Loran stations is proportional to the length of the baseline joining the pair and can be taken as approximately ·5 per cent. of range from the mid-point of the baseline.

# CHAPTER III.

## GEE.

THE Gee System operates on the general principles outlined ; that is, measurement of the time difference between the reception of a pulse signal from two transmitters and the determination of a position on a hyperbolic line from that time difference.

Gee was developed during the War and widely used by aircraft, its marine application being known at the time as " Q.H.".

The system operates on a high frequency and is therefore limited to short ranges when used as a marine aid. The presentation of the signals is visual, using a Cathode Ray Tube.

Gee differs from Loran in that the signals from two pairs of stations appear on the screen at the same time, enabling the operator to take two time-differences and these plotted on his lattice chart give him a fix. Position-fixing is thus more accurate and more speedy.

The Gee shore chain consists of a Master and two Slave stations the presentation being as shown in Fig. 6.

In the example (Fig. 6) the A. Pulse is the signal received from the Master station whilst the B. and

C. pulses are its associated Slaves. The problem of identifying the pairs of pulses constituting the chain is solved in an ingenious way as illustrated. The Master pulse which must be made to appear on the lower trace to form a pair with the C. pulse is identified by the presence of a " Ghost " pulse. The " Ghost "

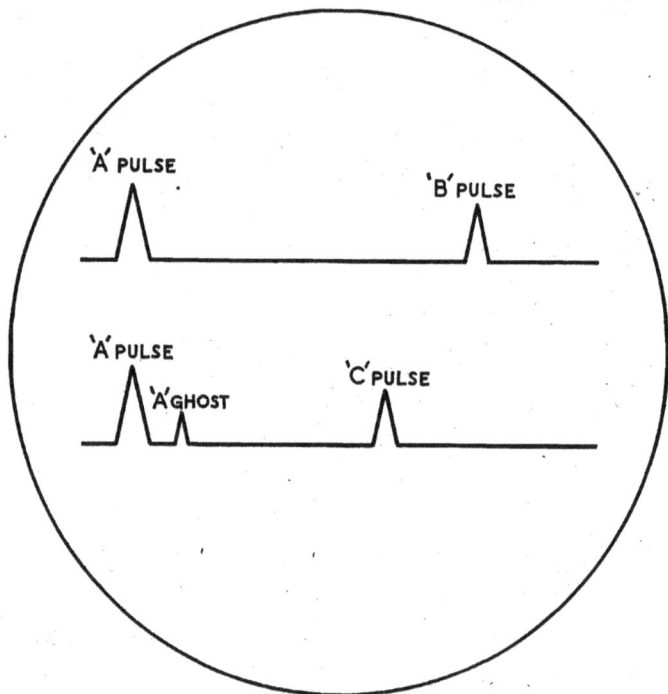

Fig. 6.

is a small pulse which follows one of the A. pulses and when lined up as shown (by means of the controls provided) it can be seen that we can obtain a time-difference A-B on the upper trace, together with a time-difference A-C on the lower trace. These two figures can then be plotted on a Gee lattice chart to give a fix.

Results show that for marine work a range of 100 miles from the shore stations is normal and an experienced operator can be expected to obtain readings of an accuracy of ·5 per cent. of range and plot his position within a minute. To obtain complete coverage over a wide area a considerable number of shore stations would be required. One of the great advantages of this system as far as the marine navigator is concerned is the fact that the indicator will provide the correct answer or none at all.

# CHAPTER IV.

## SOME OTHER SYSTEMS.

1. *The P.O.P.I. Navigation System.*—This system (the Post Office Position Indicator) has been invented by the British Post Office Engineering Department and its application to marine navigation would be as an " Ocean Aid " or a " Landfall Aid ". The technical details of both transmitters ashore, and receiver on board, are quite complicated, but the information received is in the form of bearings indicated automatically on meters. The bearings from shore beacons may be obtained up to distances of 1500 miles and, with two or more beacons sited to provide good position—line cuts over the area of operation, fixes may be obtained.

P.O.P.I. is only in the experimental stage but it is a system which offers facilities for determination of position beyond the range of normal beacon stations to an accuracy of $\pm 2°$ at up to 1500 miles.

Although employing much of the technique of the Hyperbolic Systems it differs from the true hyperbolic aids in that it displays instantaneous bearing indications on a scale and pointer meter, no lattice charts being required.

2. *Consol Radio Navigation System.*—In this system the ground station radiates Morse dot and dash

signals and a position line can be determined by making a count of the dot and dash characters heard during the transmission period of one minute.

No special equipment is required on board the receiving vessel as the usual type of medium frequency communication receiver carried will pick up the signals. A chart overprinted with the Consol lattice is required and by reference to this it is possible to establish that the vessel is located on one of five alternative position lines in each coverage sector of 120°. This ambiguity may be resolved by taking a rough D.F. bearing of the Consol station or by ordinary D.R.

The position lines are Great Circles and no special charts are required. Two ground stations are required for obtaining a fix.

The system is a long range one and provides navigational facilities up to distances of about 1500 miles over the sea and 1000 miles over the land under favourable conditions. The minimum range is 25 miles within which the system is not usable.

The operation of the system is very simple and requires only elementary training.

# CHAPTER V.

## DECCA.

THE Decca System of Hyperbolic Navigation is the system most likely to be encountered in the Merchant Navy. This is because in January, 1947, the Ministry of Transport gave official approval for its use as an aid to navigation, at least up to a radius of 250 miles from London, and this was followed by the System being unanimously approved at the International Marine Conference in New York in April, 1947. Added to these official blessings the International Telecommunications Conference in August, 1947, specially reserved wavelengths for the use of this navigational aid. In order to obtain these concessions the equipment has had to produce satisfactory results during a long period of practical operation at sea. The Decca Company are accordingly going ahead providing fuller coverage around our coasts and many more shipowners are fitting the shipborne equipment.

The Decca system is a true hyperbolic aid and a Decca Lattice chart is required in order to identify the position lines and obtain fixes. It operates on a high frequency so for the moment can be regarded as a close range aid. The principle used to determine the position line, however, differs from the systems already described, in that pulse transmissions from the shore stations are not used, the signal being sent out continuously, and that the time difference of arrival of the signals is not measured. In all other respects, as far as the navigator is concerned, the system is closely allied to the other hyperbolic aids. A special Lattice Chart is used which contains the familiar hyperbolic lines superimposed on it relative to the positions of the Master and Slave stations.

The same arrangement of shore transmitters is required, a Master station, together with three slaves giving three sets of hyperbolae, providing a complete service of position lines and fixes throughout the 360° about the stations forming a Chain. The main differences in this system are the means by which the position lines are determined and the way in which the information is displayed on board the receiving vessel.

It should be clear from the discussion on fundamentals of hyperbolic navigation that its employment depends on measurement of the *time difference* between the receiver and the two shore stations. The Decca System, however, does not measure *time* in order to get the distance difference but measures the *phase* of the radio signal.

In order to grasp the idea behind the system

let us take a simple arrangement of two stations, a Master and a Slave. These two stations transmit an identically synchronised continuous signal, and in their passage through the aether together, maintain this synchronisation, i.e., they are in step with one another or, to put it more technically, the signals are in phase. Now let us suppose these signals are intercepted and received by a ship on the equidistant line produced by the two stations. The two signals left the shore together and were in step, or in phase, when they left; having travelled the same distance at the same speed they are still in phase when they are received by a ship in this position, and if the phase of the two signals could be compared, no phase difference would be found. In the case of a ship in some other position, relative to the two shore stations, a phase difference between the two signals would be observed if these were compared. Having travelled different distances to arrive at the receiver they have arrived, as it were, out of step; providing that the signals were exactly in phase when they left the shore stations. The difference in phase will indicate the difference in distance they have travelled.

If a large number of points were plotted where the phase difference between Master and Slave signal is the same, a hyperbolic curve would be built up. We are now on familiar ground again—simply building up the hyperbolic lines about the stations in order to make our lattice chart, only instead of plotting points of constant *time difference* or *distance difference* we are using *phase difference.*

And after all as navigators we are not concerned with what is being compared, time, distance or phase, as long as our Lattice Chart is correctly drawn and our instrument compares the quantities accurately and displays them simply. This the Decometer, as the Decca people call their indicator, does.

The reader will by now be wondering why it is necessary to go to the trouble of introducing another system which compares phase (which doesn't seem to him to be as definite as time, anyway) but otherwise only seems to produce the same answer.

The reason is found in the Decca ship-borne equipment. This consists once more of a receiver and indicator. The receiver consists of a combination of highly sensitive radio receivers, one for each ground station, and a series of phase comparison circuits which compare the signals from the Master and each of the Slaves. The result of this comparison is to provide electrically the means of turning the pointer on a meter-type indicator called a Decometer (Fig. 7) to a position on the dial which is relative to the phase difference of the signals being compared. There is a separate Decometer for each pair of stations Master/Red, Master/Green, Master/ Purple.

The Decometer for the Master and Red Slave, therefore, will indicate a number, which being the result of the comparison of the Master and Red

FIG. 7.—THE DECOMETER.

*By courtesy of the Decca Navigator Co., Ltd.*

Slave transmissions, shows the position line in the Red hyperbolic pattern on which the receiver is placed at that time. Similarly so with the Green and Purple Decometers, but of course only two readings are necessary to obtain a position fix.

All the phase comparison and consequent movement of the Decometer is entirely automatic. All that the Navigator is required to do is to set the meters to their correct settings (obtained from his Lattice Chart with reference to his known position) when he leaves port and they will go on merrily indicating the changes in phase difference as his vessel moves through the hyperbolic lines. A position fix is therefore obtainable at any instant throughout the voyage without any training or specialized knowledge. All that is required is the ability to read the meters and transfer the figures obtained to the correct position line on the chart.

For convenience of reading, the lines are grouped into zones and lettered accordingly.

# CHAPTER VI.

## HYPERBOLIC NAVIGATION IN PRACTICE.

THE introduction of this aid into Merchant Ships enables the navigator to have at his elbow an accurate and automatic means of continuously fixing his position, but it is again necessary to repeat the warning that whatever instrument is used, it is still purely an aid, to be used in conjunction with his other aids plus the navigational and pilotage knowledge possessed.

This very brief account of the fundamentals of the system and the different types now in use, is put forward as a preliminary introduction to the art. The principles of operation of the ship-borne equipment are extremely simple and will be explained to the navigator when he takes charge of an installation, but this brief account of the basic principles behind the different systems will help him to a better grasp of what his instrument is doing.

Among the advantages of hyperbolic navigation is its unlimited capacity for handling ships desirous of using any particular chain. It has an immediate application to coastal shipping, as will have been realised, and for this purpose the phase comparison systems are capable of extraordinarily high accuracy.

In a coastal vessel on a regular run the navigator will soon discover that he obtains similar running fixes on successive voyages. In other words, prior to leaving port he could lay off the courses on his lattice chart and note the lattice co-ordinates, i.e., the position of a point expressed in terms of the lattice, at various points of the route. All that remains for him to do is to check that his hyperbolic navigator gives the required information at the particular points, or to adjust his course accordingly in order to make a good course and landfall.

It must be remembered, of course, that no collision-warning aid is afforded but in hazy weather, with an officer accurately plotting the ship's position by hyperbolic means, checked of course by soundings, leaving the Master to keep the bridge and clear other shipping etc., steady progress can be maintained. In such circumstances a most efficient look-out must be kept, owing to the possibility that some other ships on similar voyages may be using the same method, and even the same co-ordinates.

In places where there are irregular sets of tide or current their effect on the ship's position is immediately apparent by reference to the hyperbolic aid, and early adjustments of course can be made, in order to ensure a good run. The time saving and consequent economic advantages of hyperbolic aids being immediately obvious.

From this very brief survey of hyperbolic navigation systems, it will be seen that they are not in competition with the radar systems, but rather that they are complementary to each other. A vessel equipped with an efficient radar installation, coupled with a hyperbolic aid, would appear to have all the apparatus required for safe and speedy navigation. The former giving the navigator information with regard to obstructions in his path and movements of vessels in his vicinity, the latter fixing his own position accurately and continuously.